Understanding the Elements of the Periodic Table™

SILVER

Brian Belval

47

108

Ag

rosen
central™

The Rosen Publishing Group, Inc., New York

For Buddy and Zorro

Published in 2007 by The Rosen Publishing Group, Inc.
29 East 21st Street, New York, NY 10010

First Edition

Library of Congress Cataloging-in-Publication Data

Belval, Brian.
Silver / by Brian Belval.—1st ed.
    p. cm.—(Understanding the elements of the periodic table)
Includes bibliographical references.
ISBN 1-4042-0707-4 (library binding)
1. Silver—Popular works. 2. Periodic law—Popular works.
I. Title. II. Series.
QD181.A3B45 2007
546'.654—dc22
                                                2005034814

Manufactured in the United States of America

On the cover: Silver's square on the periodic table of elements.
Inset: Model of silver's subatomic structure.

Contents

Introduction

Two hundred years ago, there was no such thing as a photograph. If you wanted a picture of yourself or your family, you would have had to hire an artist to paint or draw a portrait. This was time-consuming and probably not very much fun. Next time you are in a museum, check out the paintings of people from the sixteenth and seventeenth centuries. To produce those pictures, the subjects had to sit relatively still for hours in their most uncomfortable clothes, so that the artists could paint their portraits.

The invention of photography in the 1830s made sitting still for hours to have a portrait done a thing of the past. The first photographs were the work of two Frenchmen, Nicéphore Niépce (1765–1833) and Louis Daguerre (1789–1851). The key to their invention was a fascinating bit of chemistry. The two men knew that silver iodide (silver combined with iodine) darkens when exposed to light. To make photographs, they coated a piece of metal with a thin layer of silver iodide. The metal with its coat, known as a plate, was placed inside a type of camera made from a box with a small hole on it. When the hole was uncovered, it would allow light to enter the camera and darken the silver iodide on the plate. The end result was a black-and-white photograph of whatever was in front of the camera.

Silver is rarely found in its pure state in nature. It is usually combined with other elements, such as sulfur, in what are known as ores. Pictured above is silver ore that has been recovered from a mine in Batopilas, Mexico.

Over time, other scientists improved upon Niépce and Daguerre's technique. One important advance was made by an American named George Eastman (1854–1932). Previously, photography was expensive and very complex. Cameras were large, the photographer needed to carry a supply of toxic chemicals, and the photographic plates were very difficult to handle. Eastman changed all that when he invented roll film. This type of film was also coated with silver iodide, but it was far easier to use. In 1888, Eastman began to sell a camera that used roll film, known as the "roll holder breast camera." It was wildly popular and introduced photography to the masses. Since then, Eastman's company, Kodak, has been the world leader in photography.

Louis Daguerre invented a photographic process known as the daguerreotype. Before the daguerreotype, it took about eight hours to capture a photographic image. For about fifteen years Daguerre and his partner, Nicéphore Niépce, worked on a way to speed up the process. In 1839, Daguerre announced the invention of the daguerreotype, a process using an iodized silver plate and requiring only thirty minutes of exposure time. Daguerreotypes must be handled carefully because they are extremely delicate and can be easily damaged when touched or exposed to moisture. At left is a daguerreotype of the inventor taken circa 1845.

More than 5,000 tons (4,536 metric tons) of silver are used each year to make film. This includes photographic film, movie film, and X-ray film (the type of film dentists and doctors use to take pictures of patients' teeth and bones). Although digital cameras, which don't require traditional film, have become popular, there are still many who prefer the older technology. Silver has been helping people take pictures for almost 200 years, and will continue to do so for many more.

Of course, silver has many more uses beyond photography and moviemaking. This book will explore the many faces of silver—and the chemistry that makes this amazing element so valuable.

Chapter One
Silver and the Periodic Table

Silver is one of the more than 100 elements that make up the universe. Silver isn't easy to find—in fact, silver is only the sixty-sixth most abundant element on Earth. Because it is rare and can be shaped into beautiful jewelry and coins, silver has been valued by different civilizations for thousands of years. In ancient Egypt and medieval Europe, silver was even more valuable than gold. That is not the case today. In the international market, gold is about sixty times more expensive than silver.

The Silver Atom

Silver, like all matter, is made out of tiny particles called atoms. Atoms themselves are made out of even smaller particles: electrons, protons, and neutrons. At the center of every atom is the nucleus. The nucleus consists of protons and neutrons packed tightly together. Protons have a positive charge, and neutrons have no charge. The nucleus serves as the solid core of the atom. A silver atom always has exactly forty-seven protons in its nucleus. This is what makes an element different from all the other elements: they have different numbers of protons. For example, the hydrogen atom has one proton in its nucleus, while the gold atom has seventy-nine.

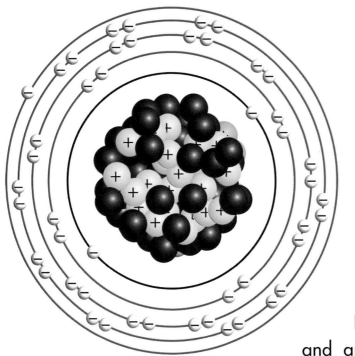

Atoms are made out of protons, neutrons, and electrons. Protons *(green)* and neutrons *(black)* are packed tightly in the nucleus at the center of the atom. Electrons *(light blue)* exist in shells outside the nucleus. A silver atom has five shells of electrons.

Electrons are negatively charged and are attracted to the positively charged protons. The attraction is similar to the way opposite poles of a magnet attract each other. Electrons are extremely small and light—they weigh about 2,000 times less than a proton or neutron. The electrons dart around outside the nucleus in paths known as shells. Each atom can have multiple shells. The silver atom has five shells of electrons. Two electrons are found in the first shell, eight in the second, eighteen in the third shell, eighteen in the fourth shell, and one in the fifth. In all the elements, and in silver, the electrons in the outer shell are known as the valence electrons. These electrons are of special interest to chemists because they allow atoms to interact with other atoms. This process is known as chemical bonding.

In each atom, the number of electrons is always equal to the number of protons. This is necessary to balance out the positive and negative charges. If an atom gains or loses an electron, it is known as an ion. An atom that has lost one or more electrons has a positive charge and is called a cation. An atom that has gained one or more electrons has a negative charge and is called an anion.

Isotopes

Atoms always have an equal number of protons and electrons (forty-seven in a silver atom), but the number of neutrons doesn't always match. Often, there are more neutrons than either protons or electrons.

Interestingly, the number of neutrons in an atom of an element can vary slightly. These atoms are known as isotopes. Silver has two isotopes: an atom with sixty and an atom with sixty-two neutrons. Scientists have calculated that 52 percent of all silver atoms have sixty neutrons, while the other 48 percent have sixty-two neutrons.

Silver isn't the only element that has isotopes. Take carbon, for example. Carbon has a famous isotope known as carbon-14. This isotope of carbon has eight neutrons and is radioactive—it breaks down over a period of thousands of years. By comparing the amount of carbon-14 in a sample to the amount of carbon-12 isotope (carbon atoms with six neutrons), scientists can figure out the age of the sample. This method is known as carbon dating.

The Periodic Table

The periodic table is designed to help you better understand the elements. In the table, the elements are arranged into rows and columns. The rows are known as periods, and the columns are known as groups. Each period and each group have a number. The periods are numbered one to seven, from top to bottom. The groups are numbered in one of two ways, depending on the version of the periodic table you are using. (Some tables, like the one on pages 40–41, number the groups using both methods.) In the first system, the groups are numbered 1 to 18, from left to right. The second numbering system uses Roman numerals (I through VIII) followed by a letter (A or B), except for the noble gases at the far right, which are in group O. In this numbering

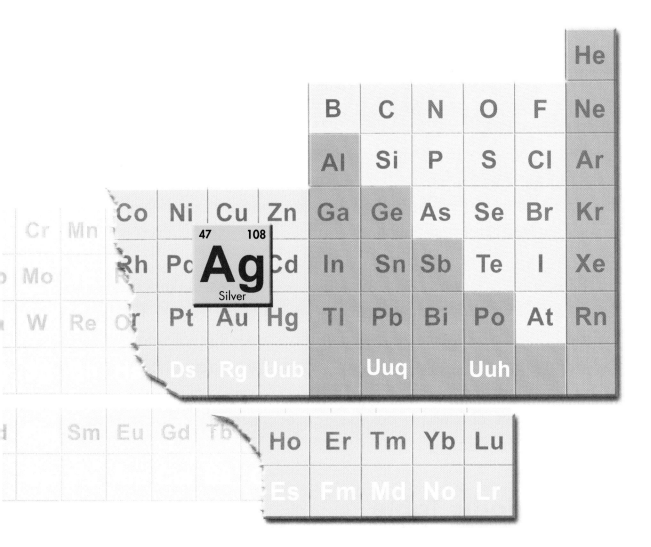

The columns of the periodic table of elements are known as groups. Silver's group includes two other well-known and valuable metals, copper and gold. Like silver, copper and gold—and all the elements in group IB—have one electron in their outer shells.

system, the elements in the groups marked with an A are known as the representative elements. In the representative elements, the Roman numeral in the group equals the number of electrons in the outer shell of an element in that group. For example, elements in group VA all have five electrons in their outer shell. Elements marked with a B are known as the transition metals. However, unlike the representative elements, the number before B in each of these groups does not always equal the number of outer-shell electrons. Like the transition metals, the noble gas elements also have varying numbers of electrons in their outer shells.

Below the main table is a smaller table consisting of two rows and fourteen columns. This block is part of the main table but has been cut out so the table isn't too wide to fit across two pages. The first row of the smaller table should go in between lanthanum and hafnium on the main table. These elements are known as the lanthanides. The second row of the smaller table goes between actinium and rutherfordium. These elements are known as the actinides.

Each element occupies one square of the table. In group 11 (IB), row 5, you will find silver. The letter or letters that represent an element are known as its chemical symbol. Silver's chemical symbol is Ag, which is short for *argentum*, silver's name in Latin. A few of the other elements

Sc	Ti	V	Cr	Mn	Fe	Co	Ni	Cu	Zn
Y	Zr	Nb	Mo	Tc	Ru	Rh	Pd	Ag	Cd
La	Hf	Ta	W	Re	Os	Ir	Pt	Au	Hg
Ac	Rf	Db	Sg	Bh	Hs	Mt	Ds	Rg	Uub

This portion of the periodic table shows the forty transition metals. These metals, which include silver, have many uses in today's world. Iron (Fe), for example, is commonly used in the construction industry. Its strength and affordability make it ideal for building large structures such as bridges and skyscrapers.

also have Latin abbreviations, such as gold (Au for *aurum*) and iron (Fe for *ferrum*).

In each square of the periodic table are two numbers. The smaller number on the top left is called the atomic number. It is equal to the number of protons in an atom of the element. As you can see, the atomic number increases from left to right and as you go down the table. The larger number on the top right of each square is the atomic weight. The atomic weight is the sum of the number of protons and the average number of neutrons in an atom of the element.

The atomic weight and atomic number can be used to estimate the number of neutrons in an atom of a specific element. This calculation isn't exact, but it gives you a good estimate of the number of neutrons in an atom of a specific element:

atomic weight − atomic number = average number of neutrons

In the case of silver:

107.868 − 47 = 60.868 neutrons

According to this formula, an atom of silver has 60.868 neutrons. However, an atom can only consist of whole numbers of neutrons, not fractions. Therefore, the average number of neutrons in a silver atom is sixty-one. In reality, as noted earlier, an atom of silver has either sixty or sixty-two neutrons. This formula can only approximate the number of neutrons.

Using the Table

The periodic table contains lots of useful information. The numbers on the table allow you to calculate the exact number of protons and electrons in an atom of an element, and also the approximate number of neutrons.

Silver Snapshot

Chemical Symbol:	Ag
Classification:	Transition metal
Properties:	Conducts electricity, malleable, ductile
Discovered by:	Has been known since prehistoric times
Atomic Number:	47
Atomic Weight:	108 atomic mass units (amu)
Protons:	47
Electrons:	47
Neutrons:	60 or 62
Density at 68°F (20°C):	10.5 grams/cubic centimeters
Melting Point:	1,763°F (962°C)
Boiling Point:	4,014°F (2,212°C)
Commonly Found:	Earth's crust in silver ores such as argentite

You can also learn a lot about an element simply from its location on the table. That is because the elements in a group often share chemical properties. (A chemical property has to do with how the element reacts with other substances.) For example, all the elements in column IA react with water, releasing bubbles of hydrogen gas.

Silver is part of a large group spanning the middle of the table. These elements are known as the transition metals. The elements of group IB are called "coinage metals" because they have been used in coins and are not very reactive (they don't corrode as iron does). The transition metals also have similar traits, including the ability to conduct heat and electricity, the ability to be pounded into sheets, and the ability to be stretched into wires. We will talk about these traits in more detail in the next chapter.

Chemists identify elements by their physical and chemical properties. In the last chapter, we briefly discussed chemical properties. This type of property has to do with how elements react with other elements or substances. These reactions result in an element being changed into something else. Physical properties, on the other hand, can be observed without reacting or changing an element into other substances. These properties include color, hardness, melting point, and density. In this chapter, we'll take a closer look at silver and its physical properties.

Phase at Room Temperature

Matter exists in three phases: solid, liquid, and gas. You are already familiar with these three phases. Think about water. Water in its liquid phase is a refreshing drink. Water in its solid phase is ice, and water in its gas phase is steam. These are the three phases of the substance with the chemical formula H_2O (two parts hydrogen to one part oxygen).

A simple way to describe a substance is by its phase at room temperature, which is approximately 68 degrees Fahrenheit (20 degrees Celsius). Water's phase at room temperature is liquid. Silver, like all the metals except mercury, is a solid at room temperature.

This hand mirror demonstrates two important properties of silver. The metal naturally reflects light and when polished makes an excellent surface for a mirror. Silver is also soft enough to be worked into intricate designs, as with the handle of this mirror. Because of this property, skilled artisans often use silver to sculpt beautiful handicrafts.

Color

It might seem obvious, but one notable property of silver is its color. Silver has a shiny gray-white appearance. You might even say that silver has a silver color! This is in contrast to the color of some other metals—such as reddish copper and yellowish gold. When polished, silver reflects light better than any other metal. Because of this property, silver is used to make high-quality mirrors.

Malleability and Ductility

Malleability is the ability to be pounded into sheets. Silver is one of the most malleable of all the metals. This property enables a piece of silver to be beaten into a sheet 0.0025 inches (0.0635 millimeters) thick. These thin sheets of silver are known as leaves. Its malleability is also the reason silver can be shaped into intricate pieces of jewelry and silverware.

Ductility is the ability to be stretched into wires. Silver is impressively ductile. Less than a half of an ounce of the metal can be stretched into a wire more than 1 mile (1.6 kilometers) long. Metals that are ductile are often used to make electrical wire. Copper is most commonly used for this purpose, mainly because it is inexpensive.

Hardness

Silver is a relatively soft metal. Because it is so soft, silver is often combined with other metals to harden it. These mixtures of metals are known as alloys. A well-known silver alloy is sterling silver, which is 92.5 percent silver and 7.5 percent copper. The addition of copper makes the metal less likely to bend or become misshapen. This is especially important in

For centuries, cultures all over the world have admired the beauty of silver and have used it in crafts. The silver buckle pictured above is from the country Nepal. Objects such as this one are highly valued for both their artisanship and their silver content.

the manufacture of products like silverware. A pure silver knife or fork would be too soft to be used as an eating utensil.

Scientists measure hardness using the Mohs scale. The scale ranges from one to ten, with one representing the softest substance (talc) and ten representing the hardest (diamond). Silver has a value of 2.5 on the Mohs scale, which makes it slightly harder than gold, but softer than most of the other metals.

Conductivity of Heat and Electricity

Electricity is the flow of electrons through a substance. Silver conducts electricity because its sole outer-shell electron can easily move from atom to atom. When voltage is applied, the electrons are set in motion, creating an electrical current. Silver is a better conductor of electricity than any other element. This property makes it very valuable in the electronics and electrical industries. In chapter 5, we will look at exactly how silver is used in these industries.

In nearly the same way that silver conducts electricity, it will also conduct heat. If you were to stir a cup of hot cocoa with a pure silver spoon, you can feel the spoon handle become hot very quickly, much more quickly than with a steel or plastic spoon. Like its ability to conduct electricity, silver is better than all the other elements at conducting heat.

Melting and Boiling Point

The melting point is the temperature at which a substance changes from solid to liquid. Silver's melting point is 1,763°F (962°C). That is much hotter than your oven at home can get. This property also makes silver a valuable element. Silver can be used at high temperatures and still retain its strength and structural integrity. Boiling point is the temperature at which a substance changes from liquid to gas. Silver has an impressive boiling point of 4,014°F (2,212°C).

Properties of Different Metals

	Silver	Gold	Aluminum	Zinc
Conducts electricity	yes	yes	yes	yes
Density (g/cm³)	10.5	19.3	2.7	7.1
Melting point	1,763°F (962°C)	1,948°F (1,064°C)	1,221°F (660°C)	787°F (420°C)
Boiling point	4,014°F (2,212°C)	5,173°F (2,856°C)	4,566°F (2,519°C)	1,665°F (907°C)

Density

Density is the measure of mass per volume. It is usually measured in grams per cubic centimeter (g/cm³). Something that is dense is compact— a lot of mass is packed into a small space.

Silver has a density of 10.5 g/cm³. Compare that to the density of aluminum at 2.7 g/cm³. Silver is almost four times as dense as aluminum. This means that if you had piece of aluminum and a piece of silver the same size, the piece of silver would be almost four times as heavy.

Silver may seem dense compared to aluminum, but compared to gold it is not so dense at all. In fact, gold (19.3 g/cm³) is nearly twice as dense as silver. Now, that's a dense element!

Chapter Three
Discovering Silver

Silver is found in the rocks of Earth's crust. The rocks that contain silver are known as ores. An example of a silver ore is argentite, which consists of silver combined with the element sulfur in a ratio of two atoms of silver to one atom of sulfur.

Silver ores are not easy to find. Geologists estimate that for every 1 billion pounds (454 million kilograms) of rock on Earth there are only 70 pounds (32 kg) of silver. However, sometimes the metal is concentrated in deposits known as lodes or veins. These have been a popular source of silver for thousands of years. The Comstock Lode was a famous lode discovered in Nevada in 1859. It yielded hundreds of millions of dollars worth of silver over a twenty-year period. By 1880, however, most of the lode's silver had been removed and mining it was no longer profitable.

The Early History of Silver

The first silver mines were discovered in Asia Minor (modern-day Turkey) and Greece nearly 5,000 years ago. Silver from these mines was worked by skilled craftspeople into jewelry, bowls, cups, and other objects. Because silver was desired by many ancient cultures, the groups or

Argentite is a silver ore also known as acanthite or silver sulfide. It has the chemical formula Ag_2S. To extract the silver, the ore must be crushed and then treated with chemicals to remove the sulfur and other impurities.

nations that controlled the sources of silver often became very wealthy and powerful.

About 600 BC, large silver deposits were discovered near Athens, Greece. Known as the Laurium mines, their silver helped build the Greek empire. Like the Greeks, the Roman people also used silver to finance their empire. Most of their silver came from mines in Spain, over which they had gained control after winning the Second Punic War in 202 BC. With a ready source of silver, both the Greeks and the Romans began using silver coins. The ancient Greeks traded the drachma, while the Romans minted a coin called the denarius. These coins ranged in weight from one-eighth to one-seventh of an ounce (or three and a half to four grams).

The Spanish mines controlled by the Romans were an important source of silver for nearly 1,000 years. Additional large mines were discovered in Germany and eastern Europe during medieval times (AD 800–1200). The silver from these mines was traded all over Europe, Asia, and North Africa. In China, the silver was often traded for the exotic spices and silk craved by Europeans.

The desire to acquire silver has resulted in much violence and suffering throughout history. The conquistadores of sixteenth-century Spain exemplify this unfortunate human trait. At left is Amable-Paul Coutan's 1835 portrait of Francisco Pizarro, a conquistador who ventured into Peru and brutally murdered the native population in order to gain its treasures of silver and gold.

Silver in the New World

In 1492, Christopher Columbus (1451–1506) landed in the Americas and claimed land for the king of Spain. He was followed soon after by Spanish soldiers and explorers known as conquistadores. These men were driven by the desire to find silver and gold. One famous conquistador was Francisco Pizarro (circa 1475–1541). His small army conquered the Incas, the native people of Peru, and claimed their vast stores of gold and silver. Pizarro's victory, however, would soon cost him his life. In 1541, he was murdered by a rival group of Spanish soldiers who wanted a larger share of the Incan treasure.

The conquistadores discovered large silver mines throughout Central and South America. One of the most profitable was the Potosí mine in Bolivia, which shipped millions of ounces of silver back to the king of Spain. For the next 300 years, mines in Bolivia, Peru, and Mexico produced the vast majority of the world's silver.

In the mid-nineteenth century, the state of Nevada would become a hot spot for silver. Rich deposits like the Comstock Lode would draw hundreds of miners and prospecting companies seeking to strike it rich. Smaller mines were also discovered in Utah and Colorado. About this time, advances in technology made it less dangerous and easier than ever to mine silver.

Silver Production Today

According to the Silver Institute, an international association of people in the silver industry, more than 634 million ounces (18 million kg) of silver were mined in 2004. That was a 4 percent increase over the previous year. As it has been since the sixteenth century, much of the world's silver comes from Central and South America. The biggest producer of silver is Mexico, followed by Peru.

Top Ten Silver Producing Countries in 2004

	millions of ounces	millions of kilograms
1. Mexico	99.2	2.81
2. Peru	98.4	2.79
3. Australia	71.9	2.04
4. China	63.8	1.81
5. Poland	43.8	1.24
6. Chile	42.8	1.21
7. Canada	40.6	1.15
8. United States	40.2	1.14
9. Russia	37.9	1.07
10. Kazakhstan	20.6	0.584

Source: The Silver Institute

In the future, Australia will likely challenge Mexico and Peru as the leader in silver production. Australia's silver has mostly been mined in the last century. Large deposits have been discovered even in the last decade. It is possible that there are other large deposits in Australia yet to be discovered. As other countries deplete their stores of silver, the world may look to Australia to meet the continuously growing need for the precious metal.

Extracting Silver from Ores

Lodes or veins with high concentrations of silver ores, such as the Comstock Lode, were for centuries a major source of the precious metal.

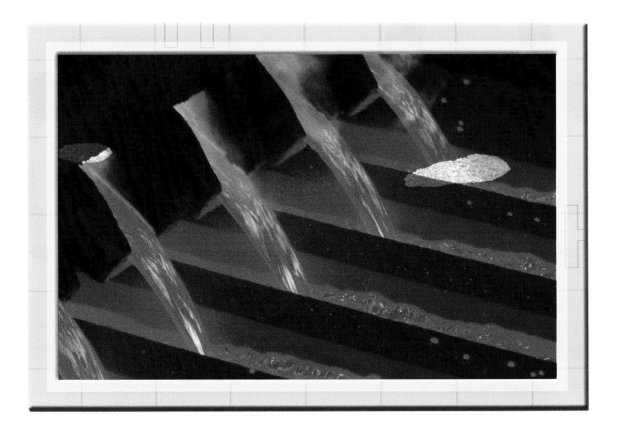

Molten copper pours from a grate at a copper processing plant in Antwerp, Belgium. This technique is known as smelting. Today, silver is often purified as a by-product of copper processing.

However, these sources have become harder to find. Today, silver is often the by-product of the processing of other metal ores, such as lead, zinc, and copper.

Galena is an example of lead ore that is mined first and foremost for its lead content but also produces small amounts of silver. The mining of galena begins in an underground mine where ore is drilled and blasted from the surrounding rock. The ore is then ground into a powder by powerful rock-crushing equipment. The separation of silver from the ore mixture involves a number of steps. The first step is known as flotation separation. In this step, the ore, water, and special chemicals are mixed in a container. The mixture is agitated to create air bubbles. Metals in the mixture attach to the air bubbles and rise to the top. The metals are then skimmed out of the container. The collection of finely powdered metals is partially melted to remove sulfur and oxygen impurities. It is then completely melted in a furnace (a process known as smelting) to remove additional impurities, such as the element antimony. The molten mixture is poured into a container of pure, molten lead that is hot at the top and cooler at the bottom. As the molten mixture is added to the container, the silver turns into crystals and floats to the surface because silver has a higher melting point than lead. This allows silver to be separated from the lead, which remains molten.

Chapter Four
Silver Compounds

Atoms of different elements join together to form compounds. For example, silver and iodine atoms combine to form a compound called silver iodide.

Why do atoms join with other atoms? The reason has to do with the valence, or outer-shell electrons, discussed in chapter 1. Chemical bonds result when atoms share or transfer valence electrons. Let's look at silver iodide as an example. Silver has one electron in its outer shell, while iodine has seven. Atoms with a filled outer shell of electrons are more stable. When an atom of iodine comes near a silver atom, it pulls away silver's lone outer electron. By gaining one electron, an iodine atom gets a filled outer shell and becomes more stable. As a result, iodine gains an electron and becomes a negative ion. By losing an electron and forming a cation, the silver atom also achieves a filled outer electron shell and becomes more stable. Silver has lost an electron and becomes a positive ion. These oppositely charged ions are attracted to each other and form a type of chemical bond known as an ionic bond. Compounds that form ionic bonds are known as ionic compounds.

Chemical Formulas and Equations

Chemical formulas are a shorthand way of describing compounds. The chemical formula indicates which atoms are bound together and in what ratio. For example, the formula for silver iodide is AgI. Ag is the chemical symbol for silver, while I is the chemical symbol for iodine. The elements combine in an atomic ratio of one to one. In other words, one atom of silver combines with one atom of iodine.

Ag_2S is the chemical formula for a compound called silver sulfide. The "2" to the right of the symbol for silver indicates that there are two atoms of silver (and one atom of sulfur) in each molecule of silver sulfide. Silver sulfide will be discussed in more detail later in this chapter.

Chemical reactions are described by chemical equations. Let's look at the reaction between silver and iodine to form silver iodide. Iiodine is represented as I_2 because it has two atoms in a molecule, and silver is represented as Ag because it's not molecular. The chemical equation for this reaction is:

$$2\ Ag\ (s)\ +\ I_2\ (g)\ \longrightarrow\ 2\ AgI\ (s)$$

Alloys

Mixtures of two or more metals are called alloys. An alloy is not a compound because the elements in the mixture do not combine in fixed ratios. In the compound silver iodide, silver and iodine atoms always combine in a ratio of two to one. However, in an alloy, the atoms of the different metals can combine in nearly any ratio. It is like mixing together different colors of paint, or mixing together nuts, raisins, and pretzels to make a snack mix.

Silver forms alloys with a number of different metals. One alloy of silver even has its own special name. Electrum is a pale yellow alloy of silver and gold that is naturally occurring. It has been known and valued since ancient times.

In all chemical equations, the substance or substances to the left of the arrow are known as the reactants, while the substance or substances to the right of the arrow are called the product. The equation shows that two atoms of silver combine with two atoms of iodine to form silver iodide. The "s" indicates that the silver and silver iodide are solids. The "g" indicates that the iodine is a gas (at room temperature, iodine is a solid).

Silver Halides

Silver halides are ionic compounds made of silver and a halogen. (A halogen is any of the elements from group VIIA of the periodic table.) The three most important silver halides are silver bromide (AgBr), silver chloride (AgCl), and the previously described silver iodide. All ionic compounds of silver are, to some degree, sensitive when exposed to light. All three of these silver compounds are used to make photographic and X-ray film. When light strikes film, it causes the silver halide to break apart into silver atoms and halogen atoms. The silver atoms, which are darker than the silver halide compound, deposit on the film. The film is then developed to bring out the image. The process of developing a silver exposure involves a chemical reaction in which silver halide is converted to silver metal. This reaction is speeded up by the tiny particles of silver formed when the film is exposed. Therefore, the developed image is darkest where the silver halide was exposed and turned into silver particles. In essence, a photograph is an image drawn with silver.

Silver iodide is also used to produce rain artificially. This procedure, known as cloud seeding, has been used in areas that are stricken by drought and in desperate need of rain.

In cloud seeding, airplanes are used to spray silver iodide into a cloud. The silver iodide causes water droplets to form into ice crystals, which fall out of the cloud. As the ice crystals fall, they melt and turn into rain.

In the experiment above, a key is placed on top of silver chloride, then exposed to light for thirty minutes (1). When the key is removed (2), the silver chloride beneath the key that was blocked from the light has not changed color, while the surrounding silver chloride has darkened from light exposure.

Silver Nitrate

Silver nitrate ($AgNO_3$) is a colorless compound that is made by reacting silver with nitric acid (HNO_3). It is poisonous and must be handled very carefully. Silver nitrate is commonly used in the preparation of silver halides for photographic film. The equation on the next page shows how silver nitrate reacts with table salt (NaCl) to create silver chloride. The "aq"

At left is an experiment that shows a precipitation reaction. First, liquid silver nitrate is poured into a flask (1). Solid potassium chloride is then added to the flask (2). This results in the precipitation of silver chloride (3), which collects on the bottom of the flask. When the mixture is exposed to light, the silver chloride begins to darken (4).

in parentheses stands for "aqueous," which means that the compounds are dissolved in water.

$$AgNO_3 \ (aq) \ + \ NaCl \ (aq) \longrightarrow AgCl \ (s) \ + \ NaNO_3 \ (aq)$$

The reaction results in solid silver chloride gathering on the bottom of the container in which the reactants are mixed. The formation of a solid when two liquids are mixed is known as a precipitation reaction.

A tarnished silver teapot is pictured beside an untarnished silver spoon. Silver tarnishes slowly when sulfur compounds in the atmosphere react with the silver surface to form black silver sulfide (Ag_2S). Commercial silver cleaners can be used to remove the layer of silver sulfide.

In the past, eye drops of 1 percent silver nitrate solution were used to prevent eye infections in newborns. This is because silver nitrate kills bacteria. Today, antibiotics, which are inexpensive and readily available, are more commonly used for this purpose.

Silver Sulfide

When an object made of silver is exposed to air long enough, it loses its shine and turns dull gray or black. This is known as tarnishing. Silver objects, such as a fork or a knife, tarnish because they react with hydrogen sulfide (H_2S) in the air to create a thin layer of silver sulfide (Ag_2S). Some foods, such as mustard and eggs, contain sulfur and can tarnish silver objects. If you use a silver knife to put mustard on your sandwich, be sure to wash it off quickly. If you don't, the knife might become coated with an unappealing layer of silver sulfide.

Argentite is silver sulfide that occurs naturally in Earth's crust. It is also known as acanthite. The Comstock Lode was a famous deposit of argentite in Nevada that was mined for its silver content.

Other Notable Silver Compounds

Besides silver halides, silver also forms numerous other compounds. Many of these have important uses in medicine and industry. For example, silver sulfadiazine ($C_{10}H_9AgN_4O_2S$) is used to treat burns, silver fulminate ($Ag_2C_2N_2O_2$) is a powerful explosive used to make fireworks, and silver oxide (Ag_2O) is used to make batteries. Like the silver halides, these compounds are ionic—they consist of positively charged silver ions combined with negatively charged ions.

Chapter Five
The Many Uses of Silver

Silver's chemical and physical properties make it a very useful element. For example, since polished silver reflects light better than any other metal, silver is used to make high-quality mirrors. The silver halides, as discussed earlier, are critical in the manufacture of photographic film.

Of course, silver is used for much more than making mirrors and film. In this chapter, we'll look at some of the other ways that silver pops up in our everyday lives.

Silver in Your Electronics

Of all the elements, silver is the best conductor of electricity. It is such a good conductor that very little silver is needed to pass an electrical current. Used in electronic equipment, a tiny bit of silver can go a long way. For example, beneath every key in a computer keyboard are two thin patches of silver separated by a tiny space, which is called a silver membrane switch. When a key is pressed, the two patches touch, sending an electrical current that races toward the computer's central processing unit. The end result is a letter appearing on the computer screen. Silver is ideal for this task because it is extremely sensitive. It only takes a light touch of the key to produce the current. Silver is also durable. You can hit the key again and again, and the

Photographic film relies on the chemistry of silver. When the film is exposed to light, silver halide crystals in the film undergo chemical changes to capture the photographic image onto the surface of the film. Later, the film is processed to make the negative useful for printing a photograph of the image.

silver contact will not fail. Silver membrane switches are used in many other machines. According to the Silver Institute, "Every time a home-owner turns on a microwave oven, dishwasher, clothes washer, or television set, the action activates a switch with silver contacts that completes the required electrical circuit."

If you have ever opened up a computer, you have probably seen a circuit board. It is the thin green piece of plastic with small circular and rectangular pieces stuck to it. On the surface of the circuit board is a mazelike pattern of lines. These devices are also found in mobile phones, electrical appliances, and televisions. Circuit boards often use

Computer circuit boards often contain small amounts of silver. Because silver is an excellent conductor of electricity, it doesn't take very much of it to pass an electrical current. In addition to computers, silver is used in many other electronic devices, such as televisions, videocassette recorders, and microwave ovens.

The Fineness of Silver

Fineness is the measure of the purity of silver. To determine fineness, multiply the percentage of silver in a sample by 10. Sterling silver, which is 92.5 percent silver, has a fineness of 925 (92.5 x 10). Less expensive silver often has a fineness near 800. The most pure type of silver is known as fine silver. It has a fineness of at least 999.

silver. A silver alloy, known as solder, is used as a glue to attach the components to the board and ensure that electricity flows freely. Because silver is such a great conductor of electricity, it takes just a tiny bit of it to get the job done.

Collecting Silver Coins

Silver was first used as a coin more than 2,500 years ago. Silver's properties make it an ideal metal for coins. Because it is malleable, it can be shaped into a thin, round coin that is easily carried. It is also soft enough that a design can be stamped on its surface. This design might identify the nation that issued the coin, and it also designates the coin's value. For much of human history, silver was the metal of choice for coins.

Today, however, all nations have removed their silver coins from circulation. This is because silver has become too expensive. The silver content in a coin, if made today, would be worth more than the face value of the coin. In the United States, silver coins were last issued in 1965. In England, silver coins were last minted by the government in 1947. Although silver coins are not used to make purchases anymore, many countries release commemorative silver coins to honor famous people,

The Kennedy half-dollar was first released in 1964, one year after United States president John F. Kennedy was assassinated in Dallas, Texas. The coin is made of 90 percent pure silver. Versions of the coin manufactured after 1964 contain significantly less silver.

places, or events. These coins are popular with collectors, and most of them will increase in value over the years.

The Many Shapes of Silver

Silver is the ideal metal for jewelry. It is soft enough to be shaped and stretched, yet durable enough to withstand regular wear and tear. It can also be polished to create a spectacular, reflective surface. Mixed with other metals to create alloys, silver's hardness and color can be customized to meet the needs of the artist.

The Price of Silver

Pure silver is bought and sold on the international silver market. Major centers for the trade of silver include New York; Chicago; London, England; Zurich, Switzerland; and Hong Kong, China. The silver market is similar to the stock market. Instead of buying stocks, people gather to bid on and purchase silver. Often, the silver is purchased by companies that use it to manufacture silver products. It is also bought by individuals as investments. Silver investors hope to sell the silver at a later date for a profit.

Just as stock prices fluctuate daily, so does the price of silver. In the last twenty-five years, silver has ranged in price from four dollars an ounce (0.03 kg) to twenty-five dollars an ounce. In 2005, the average price of silver was about seven dollars an ounce.

Forks, knives, and spoons made out of sterling silver are highly valued for their beauty. However, since silver isn't the most affordable of metals, a lot of silverware today is made of less-expensive metal. Silver-plated silverware is also popular. Silver plating is the process in which an object made out of less expensive metal is coated, or plated, with a thin layer of silver. It is done by placing the object in a solution of silver cyanide, along with a bar of silver. The object to be plated and the bar of silver are connected to a battery. This causes the silver bar to dissolve, forming silver ions to replace those that were deposited on the object, and the silver ions to flow through the solution and deposit themselves on the object.

Silver to Fill Your Teeth

Silver is used to fill cavities in teeth. The filling isn't pure silver, which would be too soft, but a mixture of silver, mercury, tin, copper, and sometimes zinc. This mixture is known as an amalgam. The amalgam is a soft paste that, when applied to a tooth, quickly hardens and expands to create a durable filling.

Silver amalgam fillings have become less popular in recent years because of concerns that the mercury in the filling is harmful to the body. However, the U.S. Food and Drug Administration and the American Dental Association have stated that they believe amalgam fillings are perfectly safe.

The Future of Silver

No doubt about it—silver has a bright future. Today, it is inside cars, cameras, cell phones, and computers. It is used to make butter knives and buttons. In the future, it may end up in the most unexpected of places. The future of silver depends on the imaginations of the artists, engineers, and chemists of the future. They are the ones who will shape it, twist it, and react it to suit the needs of an ever-changing civilization.

The Periodic Table of Elements

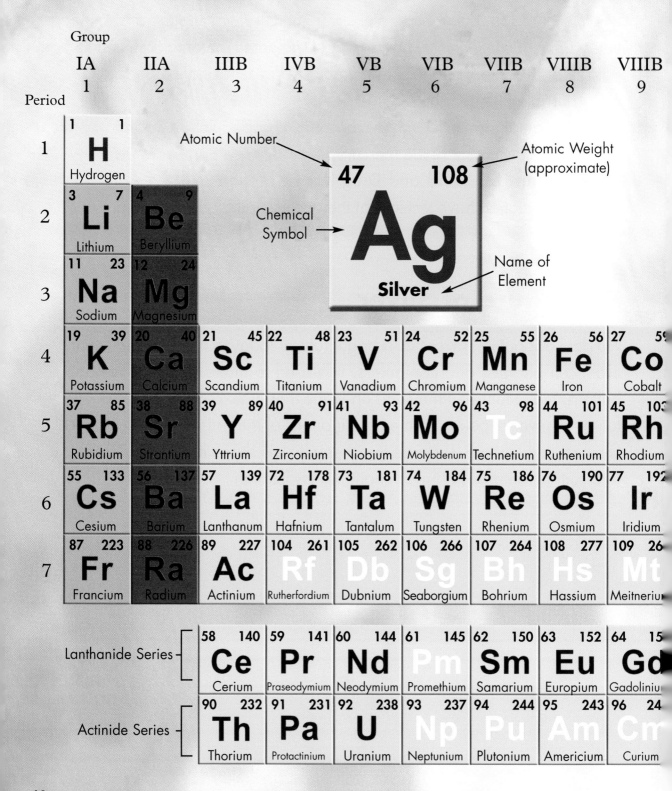

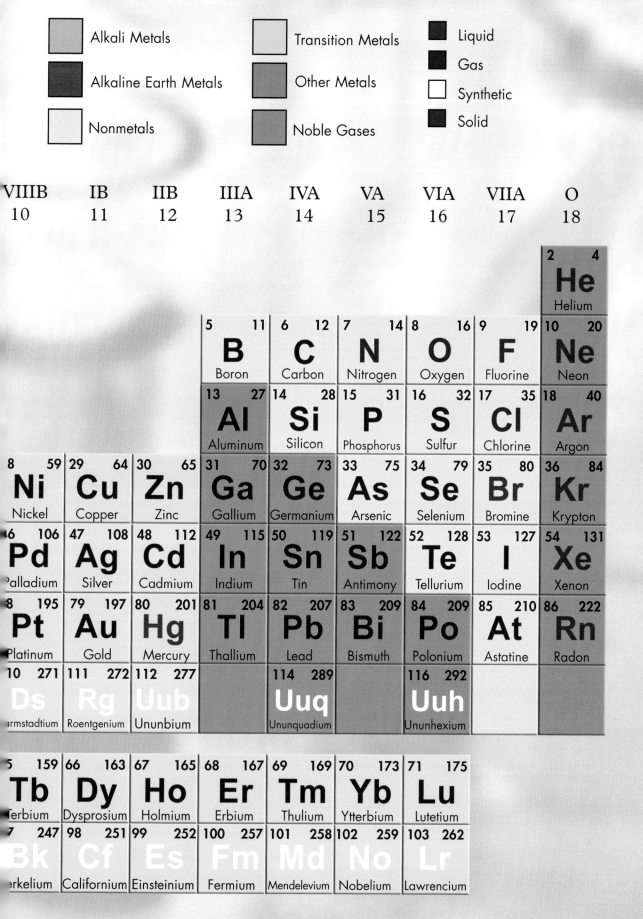

Glossary

alloy A mixture of two or more metals.

argentite A valuable silver ore, also known as silver sulfide.

atom The smallest unit of an element. The atom itself is made of electrons, protons, and neutrons.

compound A substance made up of two or more elements bound together by chemical bonds. The elements in a compound combine in fixed ratios, such as two atoms to one, or three to two, etc.

conduct To allow something, such as heat or electricity, to pass through.

current The flow of electrically charged particles.

density The mass of a sample divided by its volume.

ductile Capable of being stretched into a wire.

electron A negatively charged particle found outside of the nucleus, or center, of an atom.

fineness A measure of the purity of silver.

ion A positively or negatively charged atom or group of atoms.

lode An area that contains large amounts of metal sandwiched between layers of rock.

malleable Capable of being bent or shaped.

matter Anything that has mass and exists as a solid, liquid, or gas.

molecule Two or more atoms joined together by chemical bonds. For example, a water molecule consists of two atoms of hydrogen and one atom of oxygen chemically attached to each other. A molecule is the smallest particle of substance that still retains its chemical properties.

neutron A particle without charge that is part of the nucleus of most atoms.

ore A rock that contains a valuable metal.

precipitation The release of a substance that has been dissolved in a solution in solid form.

proton A positively charged particle that is part of the nucleus of an atom.

radioactive Capable of releasing high-energy rays or particles.

room temperature The average indoor temperature at which experiments are performed. Scientists consider room temperature to be approximately 68°F (20°C).

talc A very soft mineral also known as magnesium silicate.

tarnish Discoloration of metal due to a chemical reaction.

valence electron An electron in the outer shell of an atom.

Center for Science and Engineering Education
Lawrence Berkeley National Laboratory
1 Cyclotron Road MS 7R0222
Berkeley, CA 94720
(510) 486-5511
Web site: http://csee.lbl.gov

International Union of Pure and Applied Chemistry
IUPAC Secretariat
P.O. Box 13757
Research Triangle Park, NC 27709-3757
(919) 485-8701
Web site: http://www.iupac.org/general/FAQs/elements.html

Los Alamos National Laboratory
P.O. Box 1663
Los Alamos, NM 87545
(888) 841-8256
Web site: http://periodic.lanl.gov

Web Sites

Due to the changing nature of Internet links, the Rosen Publishing Group, Inc., has developed an online list of Web sites related to the subject of this book. This site is updated regularly. Please use this link to access the list:

http://www.rosenlinks.com/uept/silv

For Further Reading

Baldwin, Carol. *Metals*. Chicago, IL: Raintree, 2004.

Baldwin, Carol. *Mixtures, Compounds, and Solutions*. Chicago, IL: Raintree, 2004.

Heiserman, David L. *Exploring Chemical Elements and Their Compounds*. New York, NY: TAB Books, 1992.

Knapp, Brian. *Copper, Silver, and Gold*. Danbury, CT: Grolier, 2002.

Miller, Ron. *The Elements*. New York, NY: Twenty-First Century Books, 2004.

Oxlade, Chris. *Metals*. Chicago, IL: Heinemann, 2002.

Pellant, Chris. *Rocks and Minerals*. New York, NY: Dorling Kindersley, 1992.

Saunders, Nigel. *Gold and the Elements of Groups 8 to 12*. Chicago, IL: Heinemann, 2003.

Stwertka, Albert. *A Guide to the Elements*. New York, NY: Oxford University Press, 2002.

Stwertka, Albert. *The World of Atoms and Quarks*. New York, NY: Twenty-First Century Books, 1995.

Watt, Susan. *Silver*. Tarrytown, NY: Benchmark, 2003.

Wiker, Benjamin D. *The Mystery of the Periodic Table*. San Francisco, CA: Ignatius Press, 2003.

Bibliography

Atkins, P. W. *The Periodic Kingdom: A Journey into the Land of the Chemical Elements.* New York, NY: Basic Books, 1995.

Ball, Philip. *The Ingredients: A Guided Tour of the Elements.* Oxford, England: Oxford University Press, 2002.

Emsley, John. *Nature's Building Blocks: An A–Z Guide to the Elements.* Oxford, England: Oxford University Press, 2001.

Friend, J. Newton. *Man and the Chemical Elements.* New York, NY: Charles Scribner's Sons, 1961.

Geoscience Australia. "Silver Mineral Fact Sheet." Australian Atlas of Mineral Resources, Mines, and Processing Centres. 2003. Retrieved October 11, 2005 (http://www.australianminesatlas.gov.au).

Lindsay, David. "The Wizard of Photography." 1999. Retrieved October 11, 2005 (http://www.pbs.org/wgbh/amex/eastman/index.html).

Lovett, Chip, and Raymond Chang. *Understanding Chemistry.* New York, NY: McGraw-Hill, 2005.

Silver Institute. "The Indispensable Metal." Retrieved October 11, 2005 (http://www.silverinstitute.org).

Zumdahl, Steven S. *Chemistry.* Lexington, MA: D. C. Heath and Company, 1989.

Index

About the Author

Brian Belval has a bachelor's degree in biochemistry from the University of Illinois. He worked as a research scientist for a number of years before returning to school to study literature at the University of Massachusetts. He currently lives in New York City, where he combines his interest in science and writing as an editor of young-adult nonfiction.

Photo Credits

Cover, pp. 1, 8, 10, 11, 40–41 by Tahara Anderson; p. 5 © Bob Krist/Corbis; p. 6 © Jean Baptiste Sabtier/George Eastman House/Getty Images; p. 16 © B.M.W./ zefa/Corbis; p. 17 © age fotostock/Superstock; p. 21 © Stan Celestian, Glendale Community College; p. 22 © Réunion des Musées Nationaux/Art Resource, NY; p. 24 © Farrell Grehan/Corbis; pp. 29, 30 by Maura McConnell; p. 31 © James L. Amos/Photo Researchers, Inc.; p. 34 © TEK/Photo Researchers, Inc.; p. 35 © PASIEKA/Photo Researchers, Inc.; p. 37 © Fox Photos/Hulton Archive/Getty Images.

Special thanks to Megan Roberts, director of science, Region 9 Schools, New York, NY, and Jenny Ingber, high school chemistry teacher, Region 9 Schools, New York, NY, for their assistance in executing the science experiments illustrated in this book.

Designer: Tahara Anderson; Editor: Jun Lim